FOR DALIA

STEVEN M. JOHNSON

M000226340

PATENT DEPENDING
ARMBRELLA, SOFA SHOWER, UNZIPPED FLY ALARM AND OTHER ESSENTIAL PRODUCTS

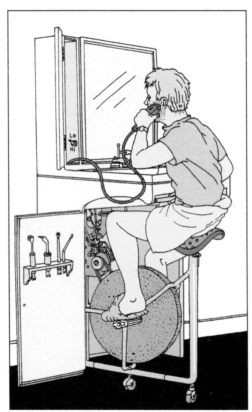

Vanity cycle with shaving attachment

PATENT DEPENDING PRESS

Also by Steven M. Johnson

What The World Needs Now: A Resource Book For Daydreamers, Frustrated Inventors, Cranks, Efficiency Experts, Utopians, Gadgeteers, Tinkerers And Just About Everybody Else
1st, 2nd and 3rd editions

Have Fun Inventing: Learn to Think up Products and Imagine Future Inventions
1st edition

Public Therapy Buses, Information Specialty Bums, Solar Cook-A-Mats and Other Visions of the 21st Century
1st, 2nd and 3rd editions

Patent Depending: Vehicles
1st edition

For information, e-mail Steven M. Johnson at jnevets@sprintmail.com.

ISBN: 978-0-692-69152-6
Patent Depending Press, Carmichael, California

www.patentdepending.com

Printed in the United States of America.

CONTENTS

What is Meant by "Essential" Products? 4

Peculiar Fashion 6

All Season Funglasses 11

Boots and Dress Shoes 12

The Efficient Employee 23

Umbrella Alternatives 34

Office Air Quality 36

Living at the Office 38

Home, Sweet Home 50

The New Bathroom 60

Consuming Food at Home 64

Addressing One's Fears 69

Being with Nature 76

Mountain Gear 81

New Types of Sports 88

Public Services 91

The Author 93

WHAT IS MEANT BY "ESSENTIAL" PRODUCTS?

Essential products are those that are unavailable for sale, that may not have been invented yet, or that exist as mere possibilities. They are not actually essential; I am merely pretending that they are essential. In 2015, I published *Patent Depending: Vehicles*, a book that dealt exclusively with products that have wheels. The products in this book are ones that have no wheels and cannot move about freely. This serves as a companion volume.

I continue to produce new work and have conceived more products with wheels, shown here: "A Different Drummer," "Drinks Door," and "Pram-shopping cart for homeless."

Over the years, I have thought up and illustrated hundreds of product concepts. The work began in 1974 as a sideline specialty in which I depicted imagined invention concepts in a cartoon format. I abandoned this sideline in 1995 but resumed it after I was "discovered" in 2009. Many of my images predicted actual future products. Back-To-School Armor on page 74 was first published

A Different Drummer—There aren't too many **A Different Drummer** showrooms yet, but they're soon to become common. The car buying public is tired of symmetrical, aerodynamically-correct automobiles. Followers of the slow car movement love these! Vehicles with ugly projections, bumpy protuberances, and bilaterally asymmetrical body styles are popular!

in 1991, and sadly it is a real product today.

A word about the use of color: My desire to show off my work in color limits the selection of works in the *Patent Depending* series to those that are already colored. On any topic, there are scores more that are still in black-and-white only.

It will be obvious to a reader that many of the intriguing or still-possible inventions and product

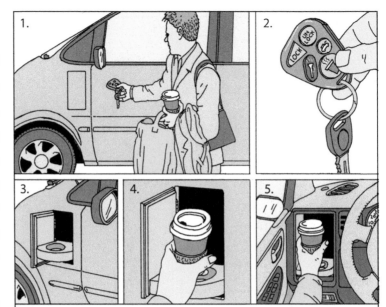

Drinks door—Unless you place a coffee cup on the roof, there's no place to put it when you open the door while carrying groceries, laptop, etc. The drinks door opens with a click of the remote. The drink is accessible inside.

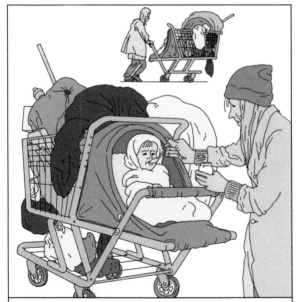

Pram-shopping cart for homeless—The number of homeless persons on the streets in America is growing. This cart offers space for baby, and for many large plastic bags.

concepts in the *Patent Depending* series were originally drawn as long as 30 or 40 years ago. This is apparent in drawings showing old-fashioned CRT computer screens, clothing popular in the 1980s, or inventions that are now outdated or irrelevant. For example, illustrations on pages 36-37 depict devices and systems for smoking indoors that would have been funny mainly during the era when offices and restaurants began to forbid smoking.

I hope you enjoy this collection.

—Steven M. Johnson, November 2016

PECULIAR FASHION

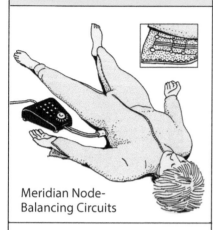

Meridian Node-Balancing Circuits

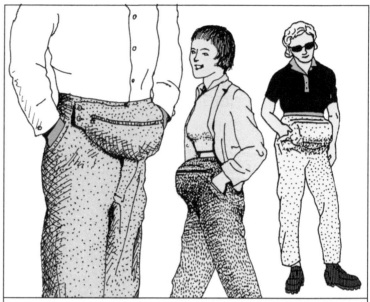

Jumpstart suit—The Meridian-Node Balancing Circuits help free up blockages in one's "energy meridians."

Pouchpants—These have a large pouch-pocket in the front. A protection from pickpockets, they afford easy access to a purse or wallet. Wearers look like human marsupials, or like they are wearing an oddly-placed codpiece.

Unzipped fly alarm—There's no word for "fear of having an unzipped fly." Scopophobia comes close: Fear of drawing attention to oneself.

Alarm

Set for shrill alarm or silent vibrate. Has 10-second delay.

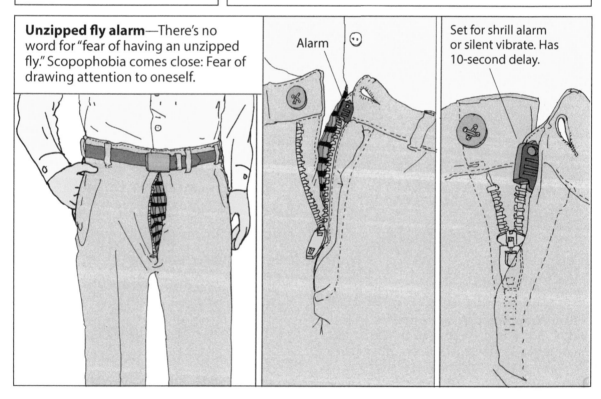

One-piece jumper

Hidden waist-high belt

Low rider-pants—Wearing pants so low, it is hard to catch a bus or escape from something or someone. These have a concealed, waist-high belt or are a one-piece jumper.

Sidewear

Ambiguity coveralls

New Orientations Wear—**Sidewear** and **Ambiguity Coveralls** allow wearers to attract extra attention, get stares and make a statement about the meaninglessness of fashion.

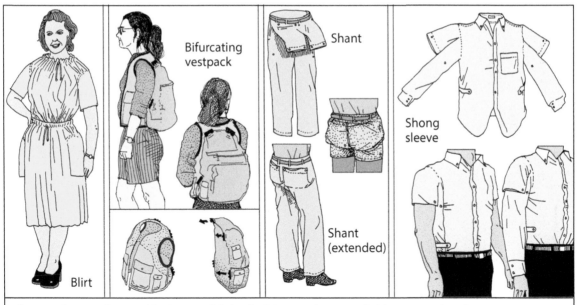

Blirt

Bifurcating vestpack

Shant

Shant (extended)

Shong sleeve

Apparel mutations—Having multiple clothing options available at any time is important. The **Blirt** is worn either as blouse or skirt. A **Bifurcating Vestpack** is worn as vest, pack or both. The **Shant** becomes either shorts or long pants. The **Shong Sleeve** shirt can be worn with long or short sleeves, depending on weather.

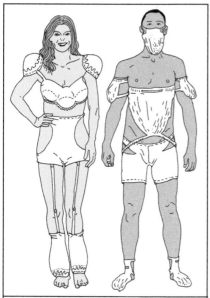

Shockingly novel undies—The old rules that required clothing to be at least practical are in the dustbin of history.

Men's underwear alternatives— These designs for so-called "new underwear" for men will shock many. Designers may choose to work with repeated themes that become the equivalent of an underwear "design language."

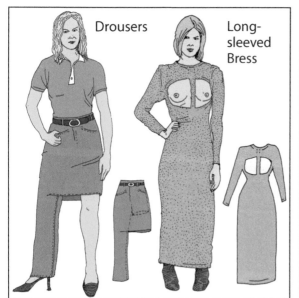

Drousers

Long-sleeved Bress

Peculiar styles, public nudity—As conventions and taboos fall worldwide, peculiar, ill-matched clothing becomes the norm, and partial or full nudity in public is common.

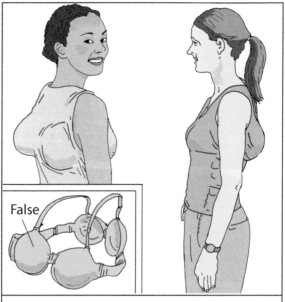

False

Doubletake brassiere—Most buyers wear this to costume parties to get a laugh. Guys think they won't be arrested for sex harassment if they only fondle the back side!

Women's underwear alternatives—When undergarments are separated from function, the designer can unleash creativity. They may ask, "What is the essence of underwearness?"

Styles ranging from foolish to ghoulish—Self-expression is important in times when it's easy to get lost in the crowd. Also, some may wish to identify with a clique in which all wear members wear sickening sunglasses.

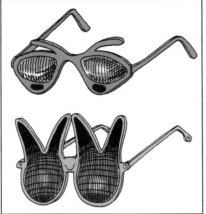

Strange eye wear—Protecting the eyes from harsh glare can now take a back seat to making a personal statement!

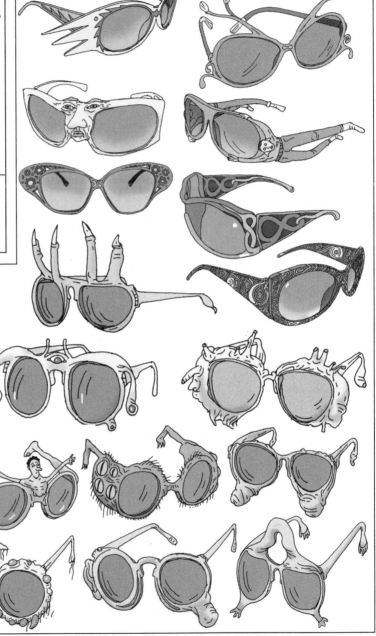

11

BOOTS AND DRESS SHOES

Pantsboots—There's need to search for shoes that go with these pants. This ensemble is a single item of clothing!

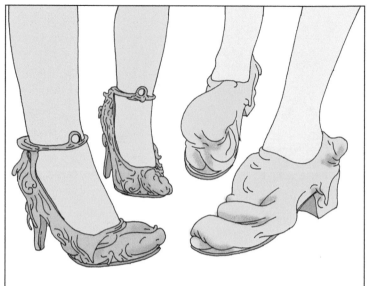

Nature's form shoes—There will surely be a demand for womens' shoes that mimic natural forms, such as stalactites, lava flows, moss, jungle vines and tide pool creatures. Nothing found in nature can be seen as ugly.

Tasteless shoes for women—Sour-tempered females may wish to own shoes that are guaranteed to be in poor taste and to offend everyone's idea of propriety.

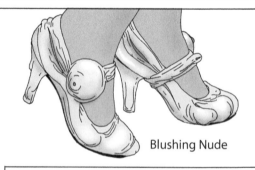

Blushing Nude

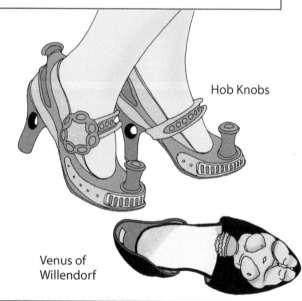

Hob Knobs

Venus of Willendorf

Pink Toe Bumper

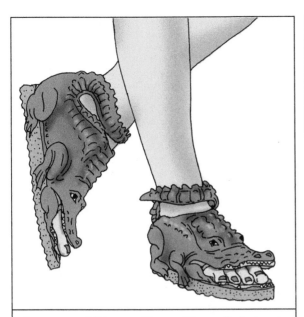

Open toe Alligator Flats—Those little teeth don't protrude; they just look nasty! The message here is subtle: I have a quirky sense of humor, but don't mess with me!

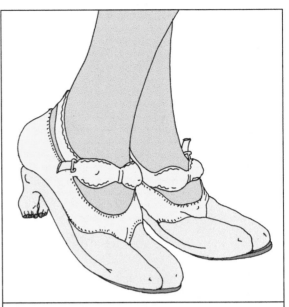

Pantie heels—These womens' shoes can be interpreted as suggestive, or as evidence the girl has an over-the-top sense of humor. Most likely no "normal" female would wear these.

Fishing for compliments—This pair of open toe high heels can easily become the talk of a party. They are beautiful and yet at the same time quite whimsical.

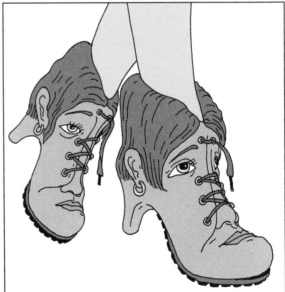

Bad mood boots—A woman may not be in the mood to attract anyone, male or female. These boots send the message that she has no desire to flirt or engage in conversation!

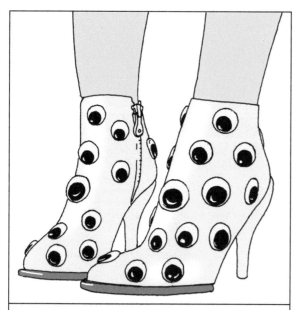

Googly Eyes heels—Not only are these heels attractive, but the pupils in the **Googly Eyes** move as you walk, making a tiny clicking noise. This is a fun pair of heels!

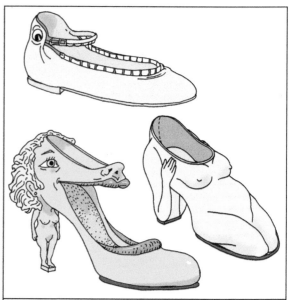

The woman shoe—The **Woman Shoe** calls attention to characteristics of females. There are smiling low heels, nudity heels, and heels that seem talkative.

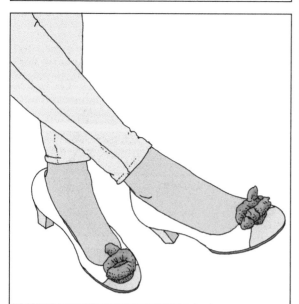

Party Pooper low heels—The decorative dog poop feature of these heels always gets a laugh. They're in the **Attention Deterred** (pronounced "de turd") line of heels.

Lumpen proletariat—There is no unique purpose or extra function served by the lumpy surface of these boots. They are popular because of this meaningless feature.

Tasteless, bizarre womens' shoes—Peculiarity for peculiarity's sake is the hallmark of these shoes. Designs may include random knobs, brackets, handles, bells or towers.

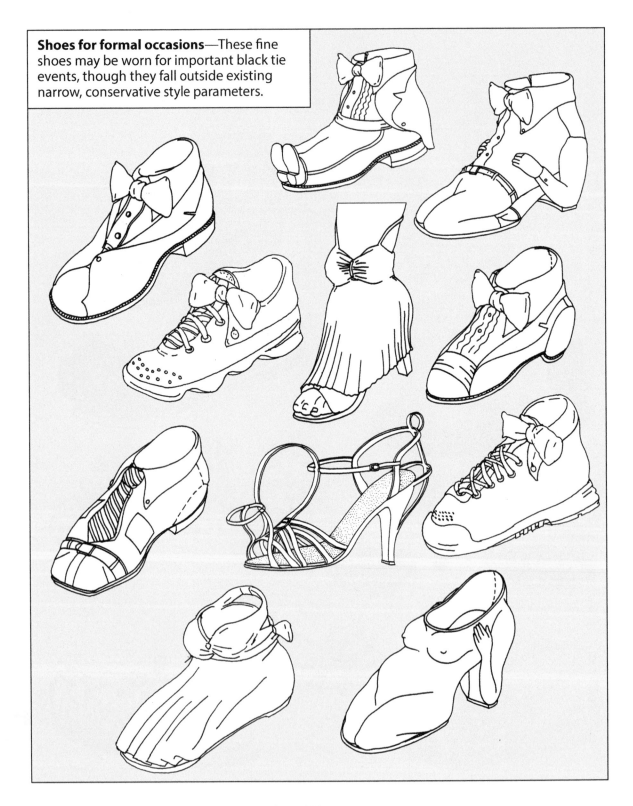

Shoes for formal occasions—These fine shoes may be worn for important black tie events, though they fall outside existing narrow, conservative style parameters.

Wing trips—The wing tip or brogue style has an uneven history as regards popularity. At times, the style falls out of favor, but then is revived for its nostalgic look.

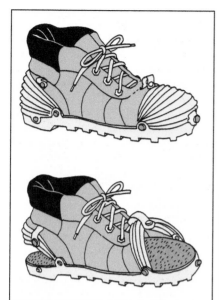

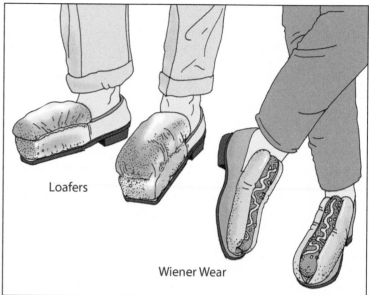

Loafers

Wiener Wear

Open toe boot—These weather-adaptable boots are not suited for stream fording or rainy day wear. They are funky.

Warm and toasties—After a long day at the office, or hours spent on the ski slopes, a guy wants to kick back and put on a pair of **Loafers.** or slip on **Wiener Wear**, especially during winter. Women love a guy who wears these!

Tasteful shoes—They come in a wide range of food-themed styles, including **Dessert Boots** and the very popular **Pizza Tops** (mushroom-olive is shown).

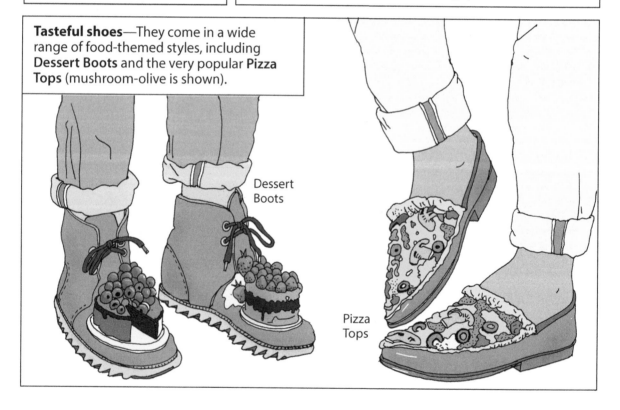

Dessert Boots

Pizza Tops

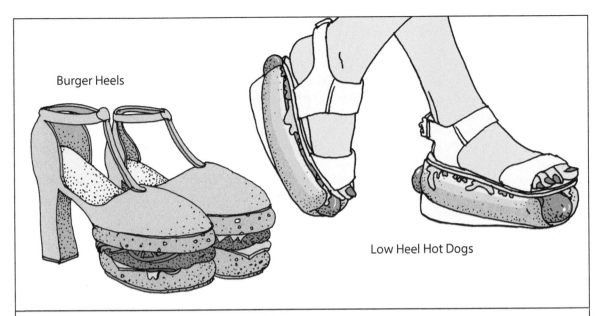

Burger Heels

Low Heel Hot Dogs

Meat eater heels—Meat eaters are considered more prone to aggressive behavior than are vegetarians. These heels let everyone know that you absolutely love to eat meat, and that you will arm wrestle any weak-muscled vegan guy or gal anytime.

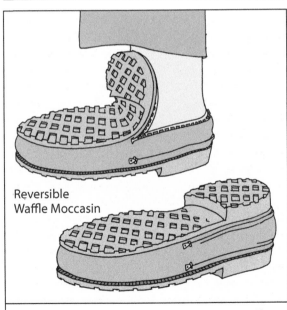

Reversible
Waffle Moccasin

Double sole Waffle Moccasin—The **Waffle Moccasin** can be worn either side up, meaning one gets twice the wear of a normal shoe. These last seemingly forever!

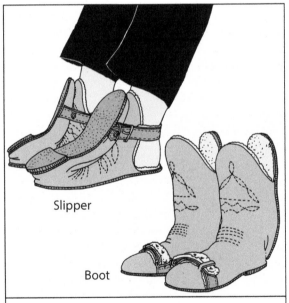

Slipper

Boot

Sheepskin boot-slipper—The boot is warmer than the slipper. You will regret wearing the slipper outdoors, particularly in rain or in a sheep, cow, or horse pasture.

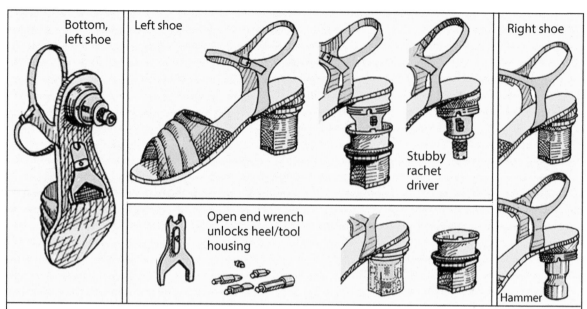

Bottom, left shoe

Left shoe

Right shoe

Stubby rachet driver

Open end wrench unlocks heel/tool housing

Hammer

High heel tool kit—it is a common belief that females are better at multi-tasking than males. If the husband or boyfriend is never around when things need fixing, she must deal with them herself. The high heel tool kit is perfect for this purpose.

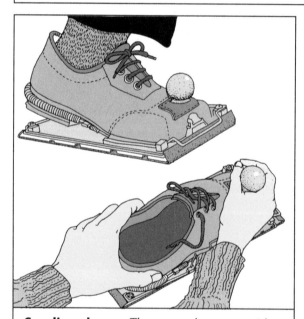

Sanding shoes—These may be worn with the sanding module attached, when there's a need to refinish floors. The module is easily removed when shoes are needed.

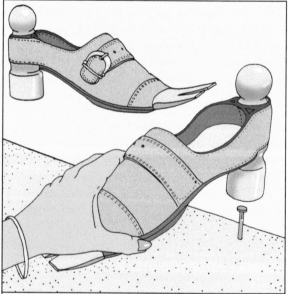

Hammer heels—When the "man of the house" is permanently stationed in front of the TV, these shoes are handy for simple carpentry tasks (but also for self-defense.)

Varieties of shoes for men—The problem with shoes for men is that the styles are not often daring, colorful, charming or outrageous. This cannot be said of women's shoes, certainly. These are funky with an attractive peculiarity that may attract some.

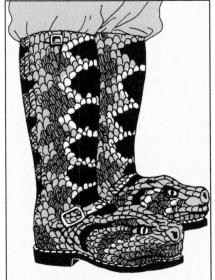

Snake charming boots—One loses the fear of encountering a rattlesnake in outdoors when these boots are worn.

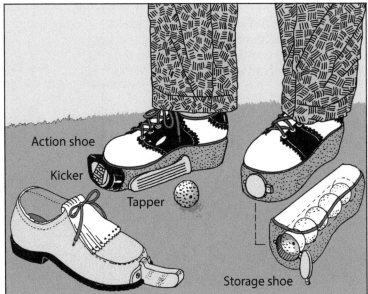

Action shoe

Kicker

Tapper

Storage shoe

Shoe Golf—They're meant for use on a **Shoe Golf Course** only. The game is never tiring, as the three-hole course is the size of four putting greens in area. When up close to the cup, use the Tapper; for distance shots, use the Kicker.

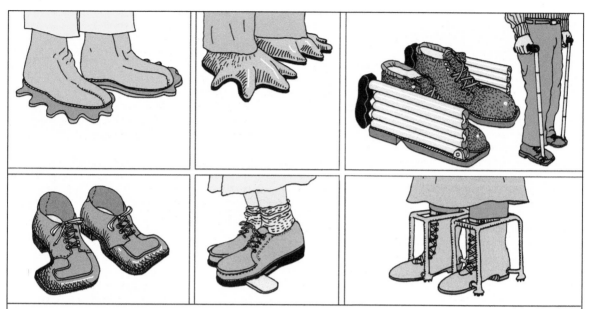

Senior stability shoes—Too much attention is paid by the media and advertising to the young, so it's easy to forget the needs of the elderly. **Senior stability shoes** may look peculiar or silly, but owning a pair can be a necessity for one with balance problems.

Tennis shoe variations—While a standard tennis shoe has identifiable components, it is possible to design shoes that echo the "idea" of a tennis shoe, but that are unique.

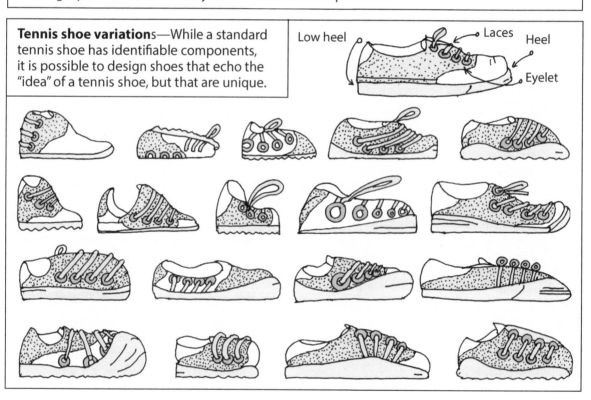

THE EFFICIENT EMPLOYEE

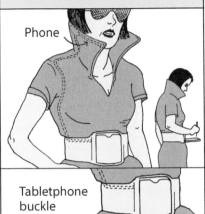

Phone

Tabletphone buckle

Tabletphone buckle—One can stay in style and also stay in touch. The earphone is hidden inside the collar!

Briefvests

Magazine Rack Vest

Magazine rack vest—information greed and time pressure may lead to anxiety over lack of time to read favorite journals and magazines. These vests allow one to easily get caught up on the latest information.

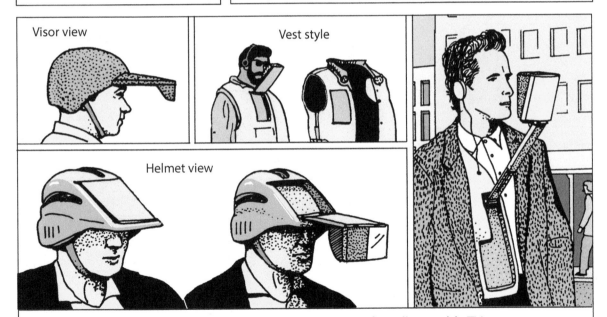

Visor view

Vest style

Helmet view

Hands-free TV and video projection wear—Designers of small, portable TVs adapt models with 2-inch screens to work inside a hat, vest, or special halter. The screen is viewed at just below eye level, allowing a wearer to keep track of looming sidewalk hazards while simultaneously watching breaking news.

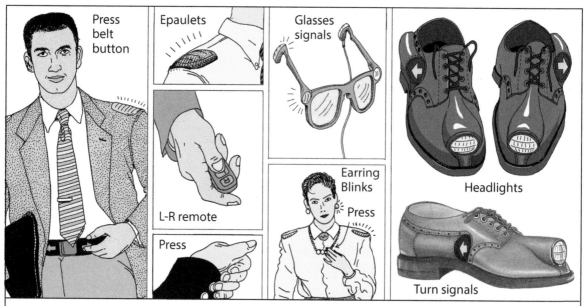

Turn signal wear—At the office, **Turn Signal Wear** helps to reduce hall-way collisions. A single blinking signal indicates one's intention to pass or turn; alternating left and right signals mean "meet for lunch"; two right signals say "I like you" : and three signals means "we're through."

Domephone—When the **Domephone** was introduced to focus groups, participants objected to having to wear a bakelite plastic hat all day. For some, it felt heavy.

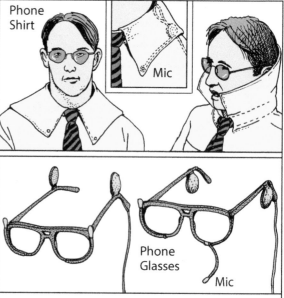

Phone shirt, phone glasses—Traditional clothing styles are modified to hold wiring and for phones inside a large shirt collar and glasses. The phone shirt is washed easily.

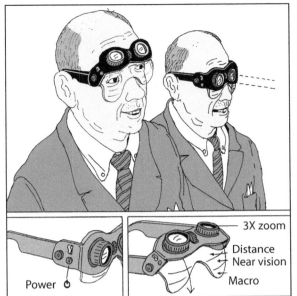

3X zoom

Distance
Near vision

Macro

Power ☉

Quadruple-focus glasses—First, they are safety glasses. They can also be matched to one's glasses prescription. Best of all, they offer four different distance settings!

Mouse overshoe

Computer mouse overshoe—The over-shoe has a hard rubber ball in the middle, and rolls on steel bearings on a pad. The opposite shoe presses a mouse clicks switch.

Camera helmets—Auto-focus cameras are fitted inside hats and helmets, with viewfinders built into attached eye wear. The shutter is depressed by touching a button on the forehead, squeezing a switch held between the teeth or in the palm of the hand, or clicking a remote. Tourists love these cameras.

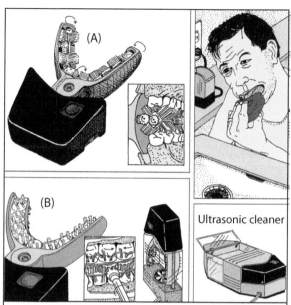

Recycle wear—Conscientious workers wear special vests or jackets with deep pockets. Aluminum can pockets are lined with removable, washable plastic liners.

Whole mouth dental appliance—One's teeth are brushed all at once with this appliance. Models (A) and (B) differ in their brushing action. A dentist fits it to your mouth.

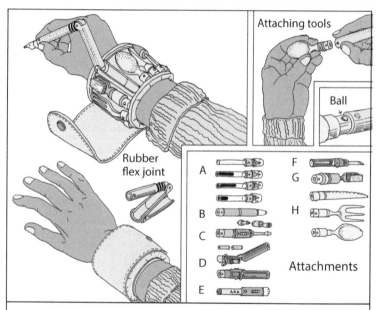

Bow Tie glasses case—It looks like a bow tie but it is a glasses case that holds sunglasses or reading glasses.

Wristband toolbox—The wristband holds tools like drafting pens (A), superglue pen (B), screwdriver (C), tear gas gun (D), flashlight (E), craft knife (F), toothbrush (G), dining tools (H), and a wide choice of other tools.

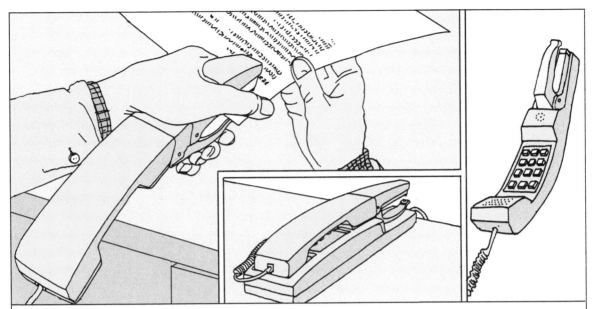

Stapler phone—It makes little sense to have many items at your desk when a single, multi-use item like the **Stapler Phone** is handy. Such a multi-use tool reduces desk clutter.

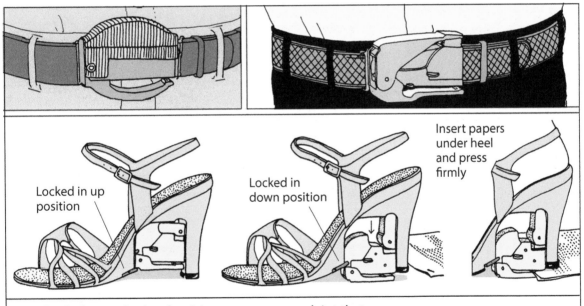

Locked in up position

Locked in down position

Insert papers under heel and press firmly

Stapler belt, stapler heels—It's a common complaint that there's a tool you want that's never there when you need it. For example, one may ask "where's my stapler?" If one wears the **Stapler Belt** all day, there's no need to look around for one.

Hands-free phone systems—Researchers have tirelessly experimented with numerous hands-free phone setups. Here are but a few samples of products they came up with.

Headset Option

Strap

Adjust phone position

Beanbag

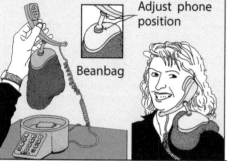

Clasp of springy steel

Forehead band

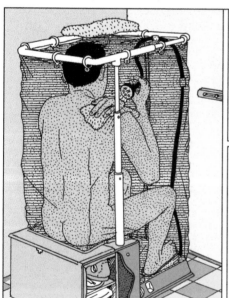

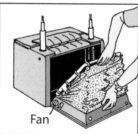

Fan

Foot pump

Fresh water

Waste

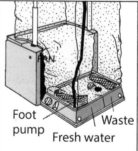

Portable gym locker—After an enthusiastic lunchtime walk or a long bike ride, an office worker who owns the portable gym locker can enjoy a warm shower while locked inside a bathroom stall. Shower water is kept warm all day inside an insulated chamber. Waste water may be dumped in a bathroom sink.

Sweat sport coat

Sweat vest

Tweed sweats

Business sweats—No one actually exercises in these suits, as they are a fad, popular at the office. They give the idea that one uses the gym and has an active, on-the-go life.

Unisex office wear—As worker roles increasingly overlap, they tend to erase gender-stereotyped job categories, and weaken the need for gender-distinct apparel.

Women's business wear—In women's business apparel, new styles may combine the provocative with the serious. Purses morph into Brief Purses, and Brief Halters.

Brief purse

Brief halter

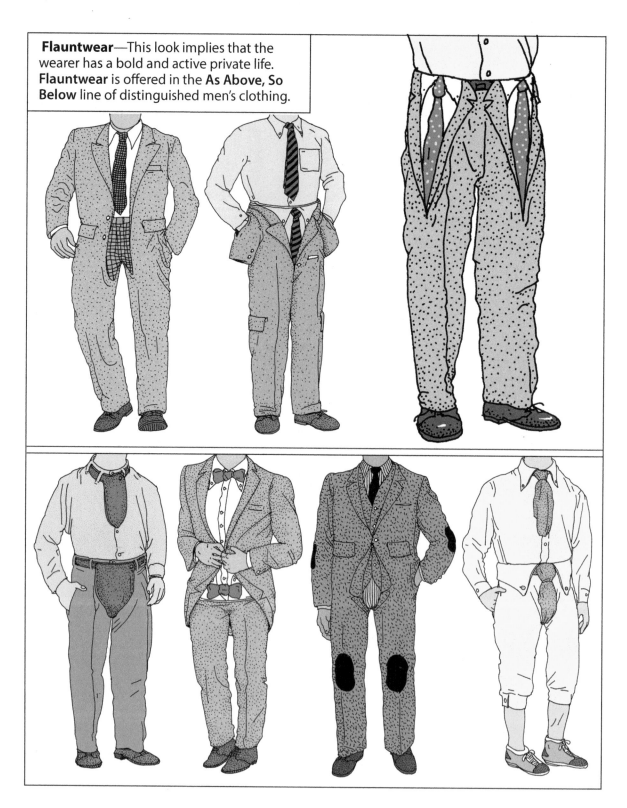

Flauntwear—This look implies that the wearer has a bold and active private life. **Flauntwear** is offered in the **As Above, So Below** line of distinguished men's clothing.

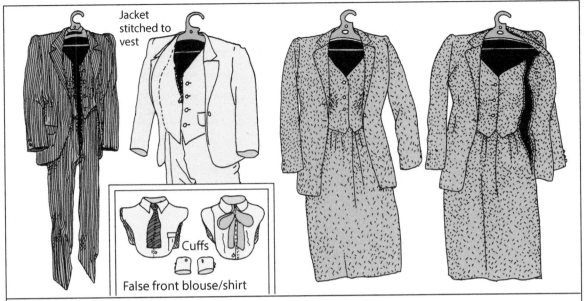

Jacket stitched to vest

Cuffs

False front blouse/shirt

Office jumpsuit–These useful outfits are designed as business-style coveralls for short-notice dinner dates as well as for emergency office meetings. They can be worn over pajamas, garden clothes, swim suit, athletic apparel or over no clothing at all!

Pre-torn officewear—In the U.S., poverty has become so prevalent that the sight of citizens wearing worn and torn clothing is increasingly commonplace. At the office, the Casual Friday tradition is morphing into full-time Poverty Chic.

UMBRELLA ALTERNATIVES

Available in black, tan or grey with matching waterproof glove.

Foam-lined shaft

Undergarment attaches to aluminum shaft.

Holstered umbrella—This umbrella is styled to look like a cap-and-ball pistol. The trigger opens the umbrella.

Armbrella—When wearing the **Armbrella**, it's not necessary to hold the umbrella; one just raises one's arm. Worn under a raincoat, the armbrella is not noticeable. A minor inconvenience in large cities is that cab drivers pull over.

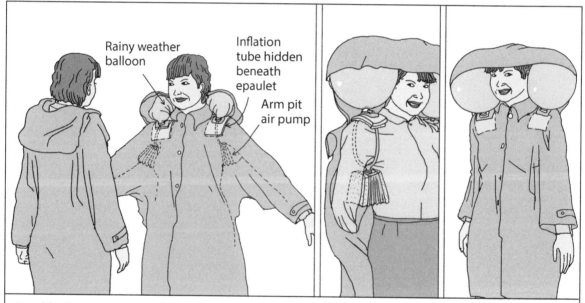

Rainy weather balloon

Inflation tube hidden beneath epaulet

Arm pit air pump

Dual balloon rain wear—Once you own stylish **Dual balloon rain wear,** you forget about ever needing to carry an umbrella. You can count on the arm pit air pump system (APAPS) to easily inflate a pair of heavy gauge latex balloons that prop up an extra-wide raincoat hood.

Tiehats in two styles.

Tiebrella

Tiehat, Tiebrella—Male (and female) office workers get in the habit in inclement weather of donning their **Tie-hat** or **Tiebrella** when heading off to work. They no longer have the excuse, "Oops, I forgot to take an umbrella."

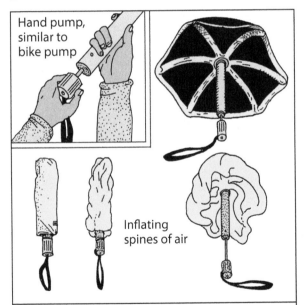

Hand pump, similar to bike pump

Inflating spines of air

Inflatable umbrella—Walking in the rain on a crowded sidewalk, one worries about being poked in the eye with an aluminum umbrella spine. This umbrella has spines of air!

Brimless

Brim Style

Head-mounted umbrella—This umbrella is out of the way, but always handy. One's hands are free for carrying packages or a briefcase, or for checking a smart phone.

OFFICE AIR QUALITY

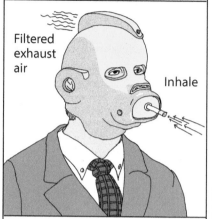

Smoker's maskhead—With its built-in "dometop" fan and filter, the **Maskhead** is perfect for indoor smoking.

To outdoors

Window breach fresh air tubes—Employee illnesses attributable to inhalation of toxic gases from wallboard, furniture glue, carpet formaldehyde, and paint VOCs force corporations to install **window breach fresh air tubes.**

Smokers' coops–Dental, medical, and auto and tire repair facilities provide **Smokers' coops** in waiting rooms. Each coop contains a super-sized activated charcoal filter.

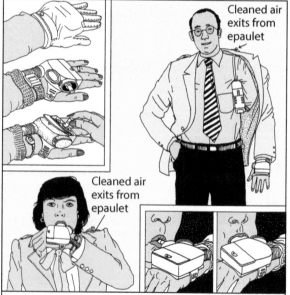

Cleaned air exits from epaulet

Cleaned air exits from epaulet

Cignull—When one wears the complicated **Cignull** smoke filtering apparatus, one gives the "signal" that one respects the right of non-smokers to breathe unpolluted air.

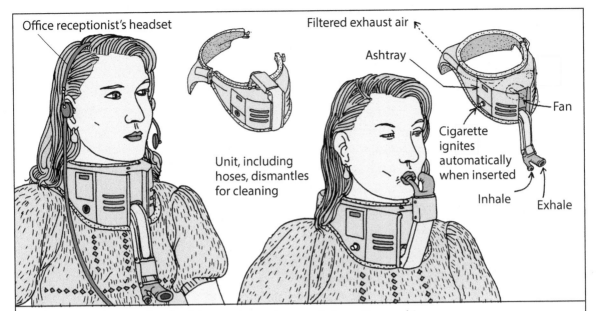

Office receptionist's headset

Filtered exhaust air

Ashtray

Fan

Cigarette ignites automatically when inserted

Unit, including hoses, dismantles for cleaning

Inhale

Exhale

Neckotine Fit—In office buildings where cigar and cigarette smoking are banned, one needs to gain permission to don a wearable smoker's filtering apparatus. The **Neckotine Fit** is a fashionable accessory that uses activated charcoal, a filter, and a fan. One's lips never touched a cigarette

Men: equally Suitable at the office or at the job construction site.

Women: Black gloss enamel, washable brim.

CIGAIRE

Cigaire—These fashionable smoking helmets have a built-in fan, filter and ashtray, and are available in several styles. They're for use by "die hard" smokers who want to smoke indoors at the office or at smoking-prohibited restaurants, while wearing their classy-looking smoking helmet.

LIVING AT THE OFFICE

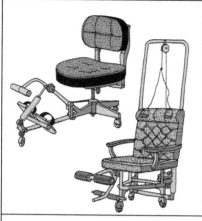

Exercise office chair—With these chairs, it is not necessary to take time off work to visit the gym. They come in many styles.

Bed above work station

Office apartment–In a competitive world market, compressed work schedules mean employees may need to stay overnight. Workmates are seen at any hour with wet hair, toothbrush in a bathrobe pocket, wearing slippers.

Office sleepers–Distinctions between work life and home life blur, as corporations adopt round-the-clock shifts to keep up with global competition. In many industries, workers are asked to sleep inside soundproof spaces under desks or in office sleepers that have a bed, change room, sink, toilet, and closet.

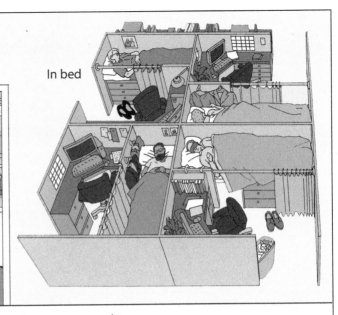

Display shows minutes left before approved nap time

In bed

NAP TIME 17MIN

Napping cubicle–Enlightened US corporations see a need to offer employees a **Napping Cubicle**. In mid-afternoon, 45 minutes are set aside for taking a siesta. Phones and computers must be turned off, lights dimmed, and talking is forbidden.

BED

BED BATH

BATH BED

The studio cubicle—The **studio cubicle** is available to employees who opt for "round-the-clock" employment. This work plan may be viewed either as a form of servitude, or as a chance for an employee to show earnestness. The cafeteria is open 24 hours for meals.

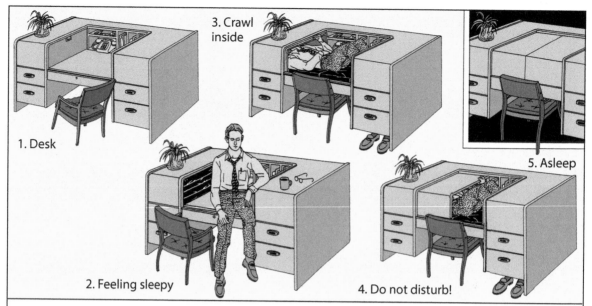

3. Crawl inside

1. Desk

2. Feeling sleepy

4. Do not disturb!

5. Asleep

Nod Office–When you begin to feel sleepy at work, the **Nod Office** comes in handy! Four futon pads are easily spread out to form a convenient bed. Insert a pair of earplugs, slide desk panels closed, turn off the phone, and fall right to sleep.

Meditating on the job

"Do not disturb"

Roofed total privacy cubicle—Not all coworkers appreciate the sound, heard especially after lunchtime, of cubicle roofs closing. The door of a roof-option cubicle can be locked and office managers know they are not to disturb an employee, however urgent the deadline.

Hiding workstations–Geniuses and shy people often dislike working in offices with open cubicles, where they are prey to aggressive type A managers and compulsive chatterers. The firm, **Introvert Interiors,** has created solitary confinement chambers (SCC) for these workers.

Luggage

Public slumber chambers—The coin-operated chamber is sanitized after each use. A sterilized pillow can be rented. There is a one-hour time limit and an alarm can be set.

Cabinet sleeper—A shallow bed with fresh air vent is included. Nappers may disturb office workers with loud snoring, or coworkers may wake sleepers if they slam drawers.

Meeting room, with occupied **office worker sleep-lockers.**

Office worker sleep-locker—When office life becomes boring or burdensome, employees may skip meetings and nap or meditate inside a sound-proofed locker. They are docked for lost work time, however.

Rotating desktops—The rectangular-shaped desktop may be inappropriate for actual work flows and task sequences. A rotating desktop offers "when-needed" access to separated heaps of reference books, take-out lunch, newspapers, bills and office equipment.

Upsy-Downsy work stations–For businesses that swing between worker layoffs and seasonal hiring binges, an **Upsy-downsy work station** is the ideal furniture choice. When employees are laid off, these small cubicles are flipped over to become reception room chairs.

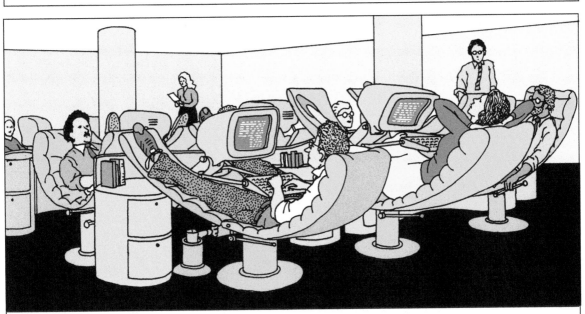

Workerlounge stations—Some information-producing offices make work life more comfortable by installing **workerlounge stations.** These create an illusion of "vacation living" ambiance, yet work quotas and worker-efficiency monitoring practices remain, preventing a sense of true relaxation.

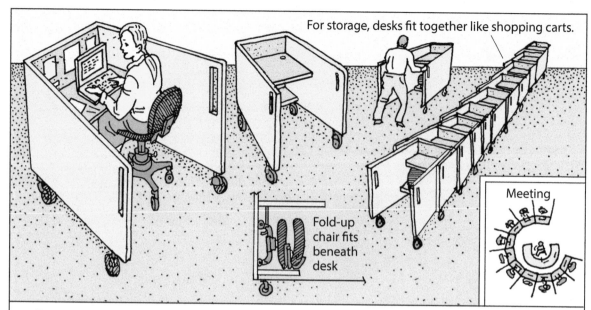

For storage, desks fit together like shopping carts.

Fold-up chair fits beneath desk

Meeting

Rollabout Desk—Rollabout desks are pushed out of the way or stored if space is needed for a yoga class, product demonstration, vendor exhibit, magic show or company gathering.

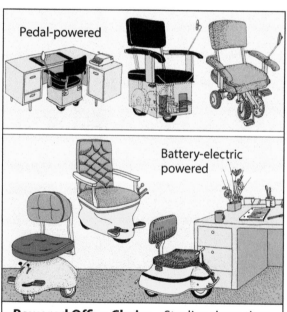

Pedal-powered

Battery-electric powered

Powered Office Chairs—Studies show that these chairs improve employee communication and efficiency, and contribute to overall worker happiness.

Deskette

Straddlefile

Roving Mini-desks—Available in **Deskette** or **Straddlefil**e models, they are ideal for impromptu meetings, taking dictation, or as a temporary desk for a new employee.

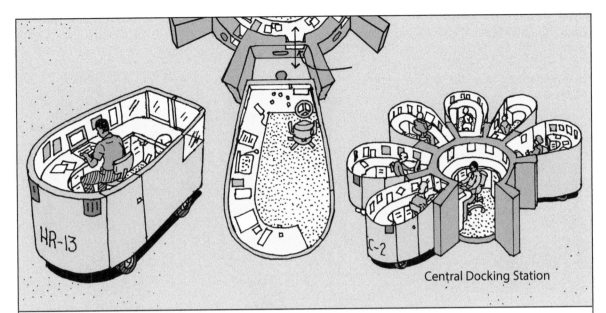

Docking Cubicle— This motorized cubicle on wheels is designed to be docked for purposes of receiving work assignments from a manager. It may also be driven to copy or coffee machines, to the bathroom, or pulled up next to another docking cubicle to discuss a joint work project, share gossip, or flirt.

Cubicle Car—Driving a car to and from work is such a strong habit these days that office workers adapt easily to spending the work day in an electric-powered roving cubicle. If there is an empty garage (green light), they may park for privacy, take a nap, or sleep overnight.

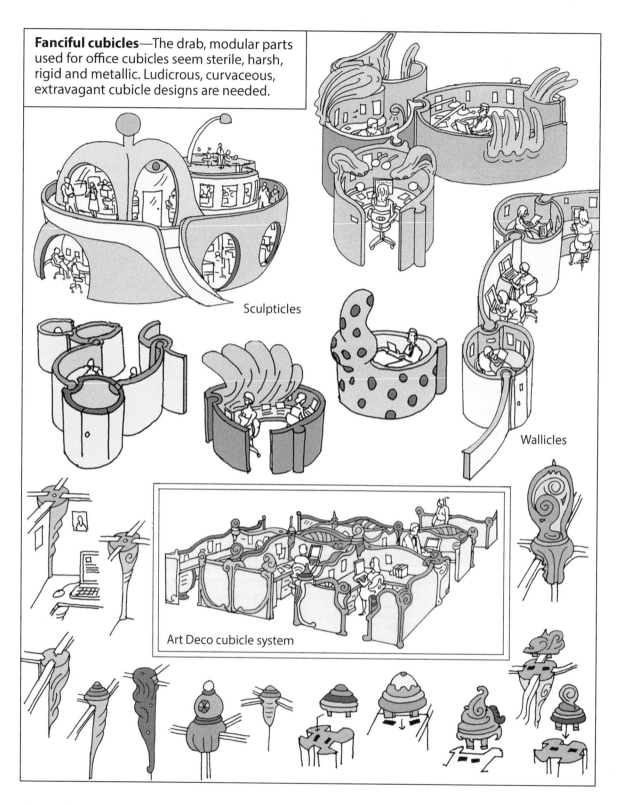

Fanciful cubicles—The drab, modular parts used for office cubicles seem sterile, harsh, rigid and metallic. Ludicrous, curvaceous, extravagant cubicle designs are needed.

Sculpticles

Wallicles

Art Deco cubicle system

Intern monitoring and prodding Station (IMPS)—It appears to be the perfect workstation, with reclining bed-chair, video-phone, and access to bill-paying services and snacks. Secretly, however, a manager hears you talk and watches your facial expression and keystrokes.

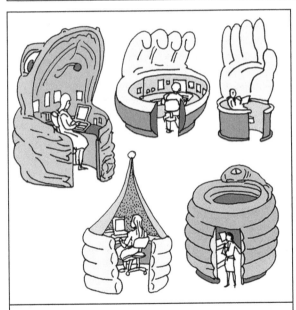

Odd work Spaces—No doubt, these spaces are unsettling to some but for others, they stimulate creativity and mental ease. Each sculptural, colorful workspace is unique.

Two-story cubicle—These workstations make efficient use of floor space. The upper cubicle has a nice view of the office and tends to be warmer than the lower space.

Squirrel cage work station–These work stations are coveted, as they save time otherwise spent at the gym. They can be noisy, and the office may smell like perspiration!

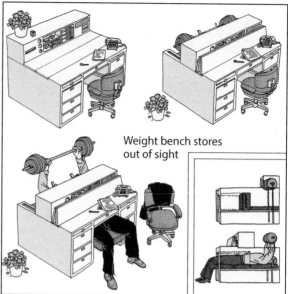

Weight bench stores out of sight

Bench press desk–It's not obvious when the desk is closed that there is a functional weight bench concealed inside. Regulation weights and lift bars are stored out of sight.

Disappearing work stations–These workstations roll over into a four-foot deep sub-floor space when the office schedule calls for yoga or tai chi classes, dance lessons, health seminars, subcontractor product expos, 401(k) seminars, or hiring freeze/layoff announcements.

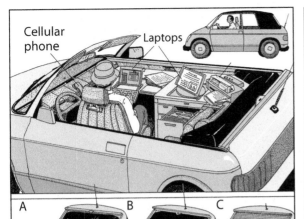

Cellular phone

Laptops

Windshield folds.

Road office–It has three configurations: Two passengers are seated (A); office work is done with a single passenger or none in the rear (B); and the shutter is closed (C).

Motorcycle office–On a nice day, take the office outdoors. Chair rotates 180 degrees. Drawers lock when motorcycle starts. Desk legs withdraw during travel. Nature beckons!

Backpack inside at back of tent

Folding table is tent roof

Portable office–These days, electronically-literate graduates of U.S. Army "learn a trade" programs, or of community college computer courses wander as itinerant, skilled bums with no place to sleep, carrying their own live-in workstations.

HOME, SWEET HOME

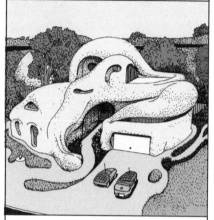

"Boneless" dwelling—there are few doors. There are no right angles. You fall or crawl into a room, roll down a hall.

Add-A-Room– After a homeowner attaches the **Add-A-Room** connector to his/her home, a tiny studio apartment can be hooked up, leveled and ready for occupation in half an hour. Rent it for extra income, or invite Granny to visit.

County Seat–This fact is not well known For little cost per month, you can rent a park bench with a secure, attached mailbox and declare it your residence and office. Living outdoors in nature has some benefits. The air is fresh and you can come and go as you wish.

Tiny Homes Urban
Development (THUD)

THUDs—As cities and counties across the U.S. give in to pressure to allow siting of mobile tiny homes on residential lots, there is no stopping the trend. Urban planners begin to urge cities to drop restrictive ordinances and replace them with THUDs.

"At times, living here feels unreal, like a dream."

"It's what we wanted. That psychologist-architect sat us at a table and made us mold shapes in clay."

Whim Homes—The plastic potential of new construction materials is usually not taken advantage of when residential subdivisions are built. Labor costs to design and build distinctive, unique structures are too high. **Whim Homes**, however, is an experiment in converting buyer fantasies into actual dwellings.

Studio Home—Young couples in the U.S. may qualify for a **Studio Home.** These homes are sited on micro-plots. They are built by manufacturers of job-site toilets.

PaySleep—A person who is tired, drunk or stoned can use **PaySleep** for $20 per night. A fresh sheet-on-a-roll moves into place, and one falls into bed as the gate closes shut.

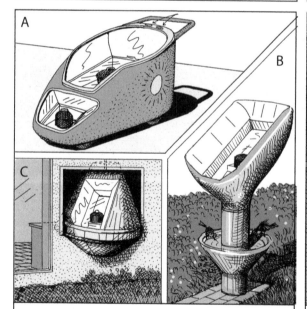

Designer sun ovens—New solar cooker styles emerge: A solar cooker on wheels (A), a self-orienting garden oven with birdbath (B), and a sun-tracking in-the-wall Solar Susan (C).

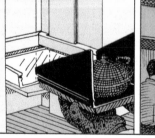

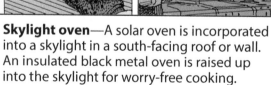

Skylight oven—A solar oven is incorporated into a skylight in a south-facing roof or wall. An insulated black metal oven is raised up into the skylight for worry-free cooking.

Lawnmowers for loafers—Demand is less these days for mower leisure furniture, as lawns are being torn up and replaced by Xeriscape gardens or organic vegetable plots.

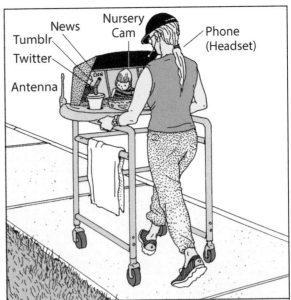

Wifi walker—Pushing a **Wifi Walker** around the neighborhood, one stays in touch with news, blogs, Facebook or Twitter, using 3G broadband service or unlocked wifi signals.

Exervac—When vacuuming, one moves a lever, choosing to engage the vacuum pump, or to move around the room.

Mother and child walkers—Mother and child move around the house, driveway or sidewalk in their matching "fold 'n store" walkers. She talks on the phone, converses on Skype, or gossips on Facebook. He uses iPad infant.

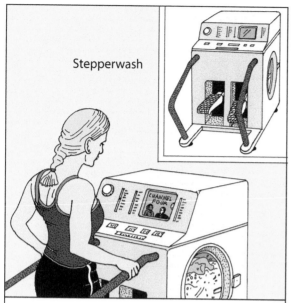

Stepperwash—The **Stepperwash** churns and cycles clothes during a 30-minute exercise session. TV, video and Internet access are available to ward off boredom.

Engine bay bed warmer packs—In winter, after arriving home from work, one pops the car hood, removes two bed warmer packs, and places them in slots under the bed.

When closed, the Pedal Wash looks like an ordinary powered washing machine.

Pedal wash—Like its cousin the **Stepperwash**, the **Pedal Wash** helps one get rid of the guilty notion that exercising at one's home or apartment is an unproductive use of time. The pedal-wash will easily do a small load in about 20 minutes.

Pedal Juicer, Exercuisine—During a juices-only fast, it is smart to get a workout making fresh, enzyme-rich raw vegetable and fruit juices. These are expensive appliances, but worth the price.

Retro-Fit bathroom appliances—People who grow up in modern industrialized societies are spoiled. They depend on gadgets that could fail at any moment if power goes out or if batteries run down. You stay fit in a home with **Retro-fit** appliances like the **Vanity Cycle**.

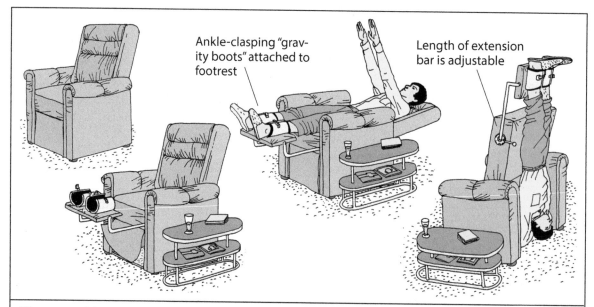

Ankle-clasping "gravity boots" attached to footrest

Length of extension bar is adjustable

Three-position incliner—A few decades ago, inversion beds were the rage. People all over America were hanging upside down, claiming that the discs in their spine were less compressed There was even the **Three-position Incliner,** hidden inside a recliner chair.

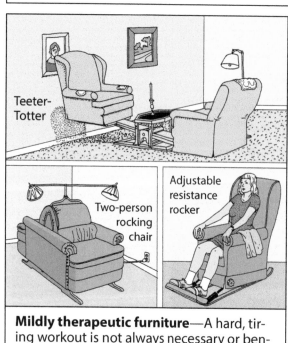

Teeter-Totter

Two-person rocking chair

Adjustable resistance rocker

Mildly therapeutic furniture—A hard, tiring workout is not always necessary or beneficial. Mild, mindless rocking or rhythmic swinging may be all the therapy one needs.

Covert gym–In a tiny home or micro-apartment, each piece of furniture may need to serve multiple functions. The **Covert Gym** is a rowing machine hidden in a lounge chair.

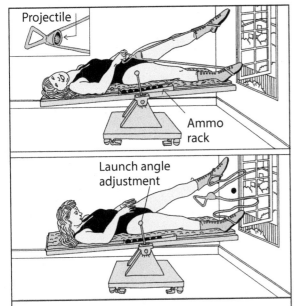

Projectile

Ammo rack

Launch angle adjustment

Exercise sling shot—This giant sling shot is therapeutic for users, but can be a problem for neighbors if it is aimed out a second floor window and large projectiles are launched.

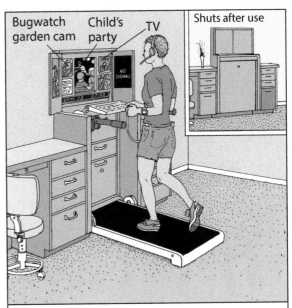

Bugwatch garden cam   Child's party   TV

Shuts after use

Infotainment treadmill—Running on a treadmill is boring, mind-sapping and tiring but time goes by fast with an **Infotainment Treadmill**. One can even type on a keyboard!

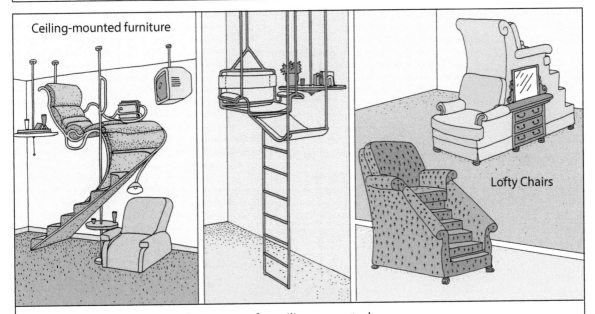

Ceiling-mounted furniture

Lofty Chairs

High End furnishings–Product names for ceiling-mounted furnishings include "Above it All" and "Lookout". One may also buy **Lofty Chairs.** Some people believe they think more clearly, or stay warmer, or are otherwise benefited by sitting up high.

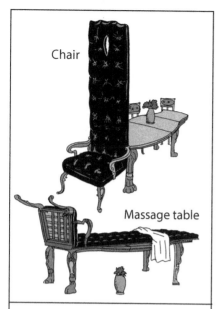

Massage table-chair—A wealthy individual may prefer stately, elegant furnishings yet need to economize on space.

Rotating room group—Multi-use furnishings make living in a small home or apartment possible. The **Rotating Room Group** is a single piece of furniture that rotates between a desk, divan, and entertainment center.

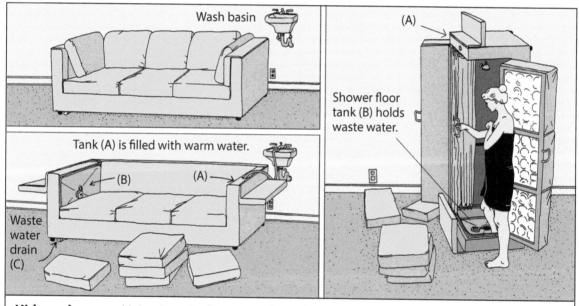

Hide-a-shower—Living in a single room or boarding house, one appreciates the easily-tipped **Hide-A-Shower.** From a wash basin tank (A) is filled with warm water. Shower floor (B) collects waste water, which is emptied from a drain (c).

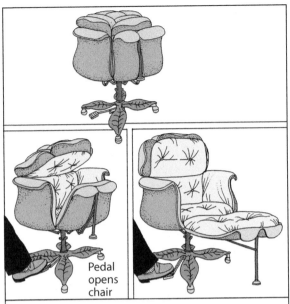

Rose chair—The **Rose Chair** is a surprising design. When closed, it looks like nothing that is familiar. When opened, it is a lovely, comfortable, cushioned chair.

Pedal opens chair

News nook—At times, couples or room-mates can't stand each other! The **News Nook** allows one to literally shut off a boring breakfast conversation, and watch the news!

Wheels

Ironing items

Ironing seat—the **Ironing seat** is an excellent TV-viewing station that's easily flipped over when a shirt needs ironing. Supplies are stored in several handy compartments.

Setting up

Inflatable guest room—This Oriental-style screen conceals inflatable furniture and a tatami mat on the back side. For a guest, it is tipped flat and the furniture is inflated!

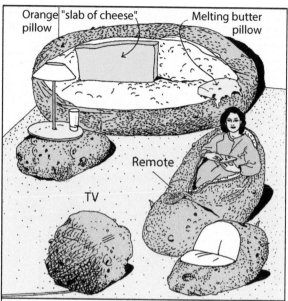

Meatless diet decor—Fanatical vegans can choose to adopt a completely plant-based apartment decor. It is not necessary for them to compromise their lifestyle or beliefs.

Potato Couch room group—Spud-themed furnishings are very comfortable and also amusing to look at. The spud comforter zips up for warmth on cold winter nights.

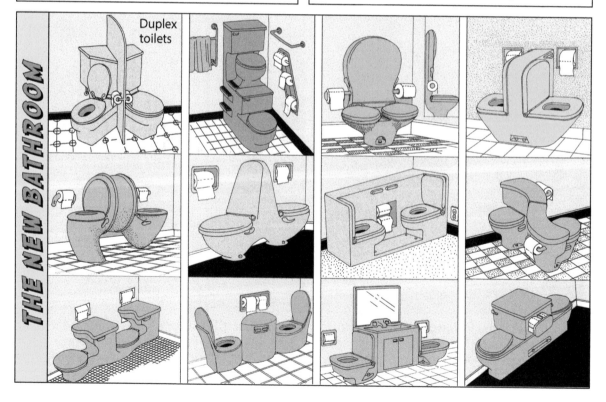

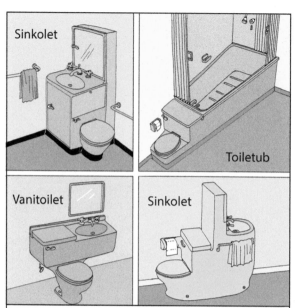

Toilet-sink combinations—When the toilet is combined with the sink or tub, interior space is gained. The bathroom becomes roomy, with more space for a wheelchair, etc.

Funk tub—The standardization of plumbing fixtures makes sense especially when there is a leak! But whimsical, one-of-a-kind tubs, handles, faucets, and knobs are possible.

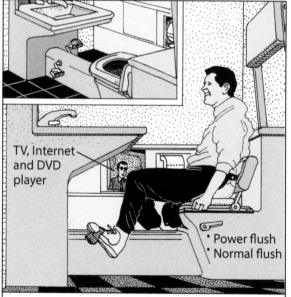

Exertoilet—Before, during, or after toilet use, one may get a workout with the **Exertoilet's** built-in exercise bicycle. RPM and miles traveled are displayed on an LCD screen.

Image-recording bathroom mirror—The mirror is also an LCD screen. It includes a video camera to record one's face over time and has replay, zoom and freeze features.

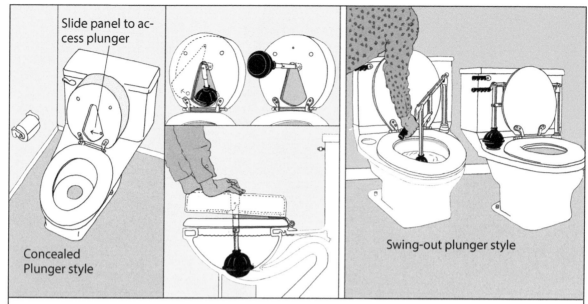

Handy Plunger toilets—When a toilet overflows, a plunger is often not within easy reach. **Handy Plunger** toilets provide a built-in plunger, useful in an emergency. The difficulty with either model is the need to wash and dry the plunger after use. However, the advantage may outweigh the disadvantage.

Communications commode—An employee who ponders a pending sale all night makes immediate use in the morning of the "facilities"–his or her **Communications Commode**. Its sophisticated desktop swivels across the lap, giving access to a phone, technical magazines or a computer bulletin board.

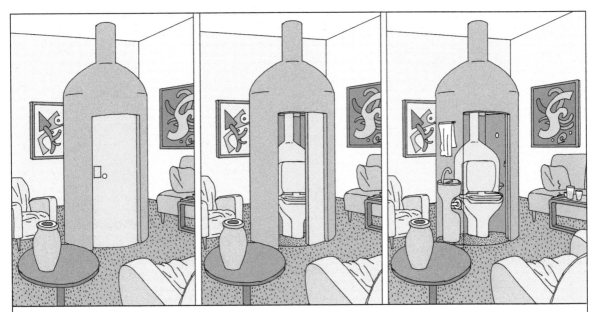

The Party Pooper—The **Party Pooper** living room toilet becomes the subject of conversation in every home where it's installed. It's so handy! After closing the door, a user may opt to press a button for "fan" or "loud music".

Living bathroom—A one-room cabin, a student's room, a studio apartment–these can be tastefully planned to incorporate a **living bathroom**, with hidden toilet, sink and medicine cabinet.

CONSUMING FOOD AT HOME

Drink-holding buckle rotates, snapping into place

Beverage buckle—The buckle conveniently frees one's hands for uses other than holding a drink, like for gesturing.

Lonely Diner

Microwave
TV
Dishwasher

Frig

Dinewash

Dinewash appliances—Why are there so many appliances and furnishings dedicated to eating? If you don't have a partner or family–maybe they moved away or died–the **Lonely Diner** or **Dinewash** appliance are all you need!

Magnets

Using the disposal

Lazy Bachelor dining system—The system combines into a single unit a polished, drop-down metal dining table, roll down shutter, dishes, place mats, disposal and dishwasher. Magnetized dishes and dinnerware are washed once the table is tipped vertically. The table seats four comfortably.

Rolling serving bowl—A "lazy Susan" is too slow for this speeded-up age. A **Rolling Serving Bowl** allows for just-in-time delivery. It should not be shoved too hard, however!

TV-watching pants—The pants have large pockets that hold VHS tape, DVD, or remote control. The snack pouch's liner is removable for washing out popcorn and drink stains.

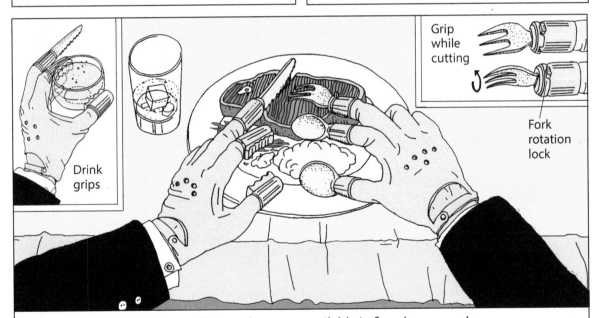

Silver Wear dining gloves—Dining gloves are available in finer homes and restaurants. The gloves offer a sanitary, scientific interface with food. If all diners wore these gloves, the typical noisy clatter of picking up and putting down dinnerware would be absent. Using them takes practice, however.

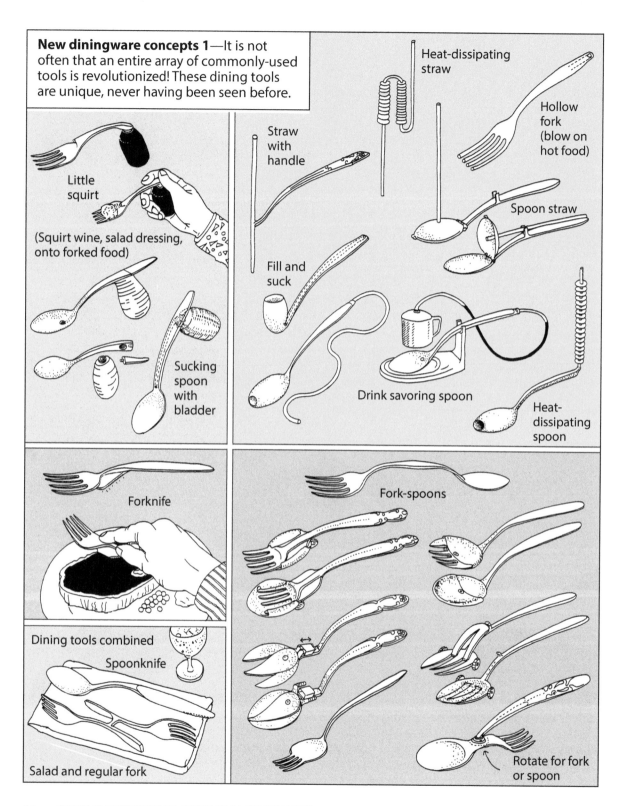

New diningware concepts 1—It is not often that an entire array of commonly-used tools is revolutionized! These dining tools are unique, never having been seen before.

Little squirt

(Squirt wine, salad dressing, onto forked food)

Sucking spoon with bladder

Heat-dissipating straw

Straw with handle

Hollow fork (blow on hot food)

Spoon straw

Fill and suck

Drink savoring spoon

Heat-dissipating spoon

Forknife

Fork-spoons

Dining tools combined

Spoonknife

Salad and regular fork

Rotate for fork or spoon

Burger grip

New diningware concepts 2—Gripping, grabbing and securing food on a slippery, smooth plate is a difficult art to master as a child (or adult). These tools lessen the stress.

No-fault chopsticks

Frankgrip

Food magnifier

Spring-loaded condiment launch

Auto-meal

Plate rotation, dinnerware stations

Rotoplate

Plate vise

Drink serving arm

Disci-plate with spikes

Slippage-minimizing surface

Rolling food arena

Steak fork

VacuDine—A food vacuuming machine is a pioneering technology that serves to bypass traditional dining steps taken to carve, capture, and cut food items. If a food piece is small enough to be sucked up inside the intake tube, a food compressor turns it into an easily digested slurry.

Glopper—This modern food-conversion cannister delivers a fine, fragrant "glop" of easily absorbed food, perfect for the discriminating diner. Food is condensed into a near liquid form, and mixed with precise amounts of appropriate food-digesting enzymes.

ADDRESSING ONE'S FEARS

Endangered person escort— It's for people who fear for their life. An armored cart is conducted into a building.

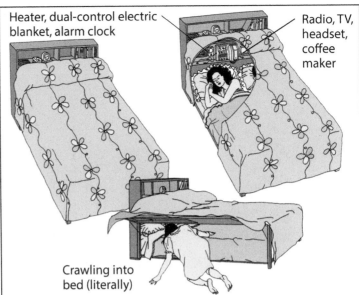

Heater, dual-control electric blanket, alarm clock

Radio, TV, headset, coffee maker

Crawling into bed (literally)

The Bed Room—As long as you don't breathe too heavily, snore, mumble, twitch or thrash around, an intruder will never suspect that there is someone hiding inside this unusual bed. Use a button or phone for summoning police.

SOUNDSLEEP
HOME BULLETPROOFING
AND INSULATION
365-9000

STOPSHOT
RESISTANCE RAT. .95

Home bullet-proofing—Gunshot sounds disturb the peace in some neighborhoods. Home bullet proofing services guarantee their window glass and in-the-wall padding.

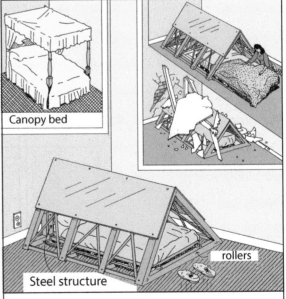

Canopy bed

rollers

Steel structure

California-style earthquake canopy bed— If you lay awake at night fearing that the ceiling will fall in during a quake, the all-steel **earthquake canopy bed** will remove the fear.

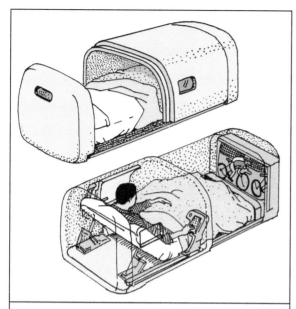

Hiding capsule—If you're a self-absorbed, unhelpful, unemployed person who stays indoors at home watching TV or Netflix rentals, the hiding capsule is perfect for you.

Recline-Insider—You can relax once you are inside your **Recline-Insider**. A home invasion gang will not know you are at home. (A police dog will discover you, however.)

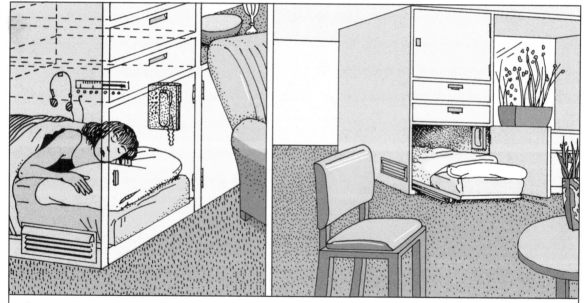

Sleeping and hiding cabinet—This concealed bed is bullet-proofed and built into a cabinet. It is also sound-proofed–lest a hiding resident's breathing should give away their location–and contains a hidden telephone with a "muffle-speak" feature that allows one to safely speak to a 911 operator.

Womb-With-a-View—When you feel disconnected, are ill, or are going through a divorce or breakup, you may be unable to appreciate friends or family. You may decide to rent a personal cave called **Womb-With-a-View**. You are free to curl up, feel sorry for yourself, and take stock.

Fear furniture sales—Fear furniture is increasingly popular. Prospective buyers see displays of false panels, escape hatches, hidden corridors and secret entries. Some furniture is bullet-proofed and sound-proofed. **Worst Fears Rooms** show off extreme personal concealment systems.

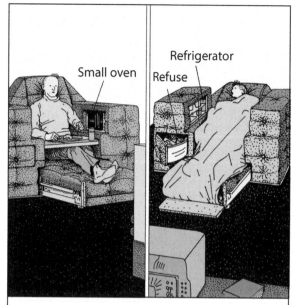

TV life support system—It is not necessary to get a job, or even to leave the house with the **TV Life Support System.** You can sleep your life away spending time in this chair.

Underground bedroom—This secret bedroom is accessed by crawling into a fake, front-loading dryer and sliding down a ramp. To exit, do the steps in reverse order.

Underground community—Shy, secretive, suspicious Americans who live in fear of–or are in trouble with–the law, may choose to change their name and live underground. These single and multi-story dwellings are cool, quiet and private.

Swimming Moat—In hard-hit drought areas of the Western U.S., swimming pools are less popular, mostly due to water use restrictions. Though new pool construction is down, home security firms report that the **swimming moat** is a hot seller.

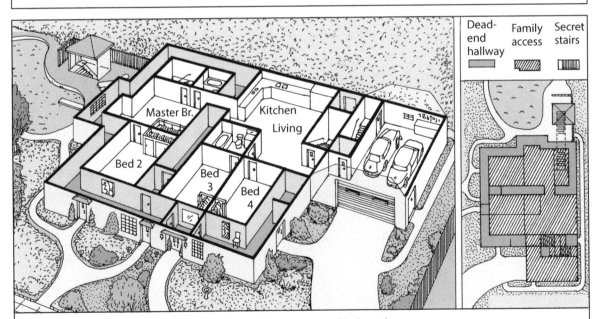

Dead-end hallway	Family access	Secret stairs

Home-invasion-thwarting mansion—Extreme wealth has disadvantages such as the threat of home invasion. This mansion is designed with a labyrinth floor plan designed to confuse and discourage robbers. Hallways shown in red go nowhere.

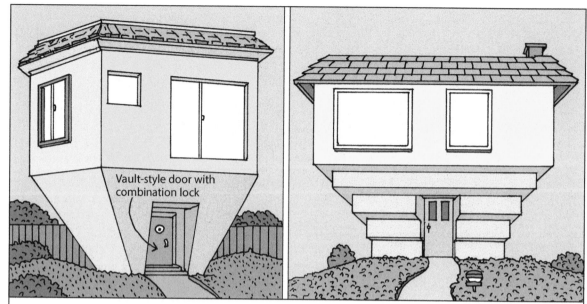

Peer-amid Homes—The **Peer-amid** home is built like an inverted pyramid. Occupants can look down on approaching sales people, religious proselytizers, or criminals, and can peer "amid" neighboring homes. It is impossible for an intruder to scale this building.

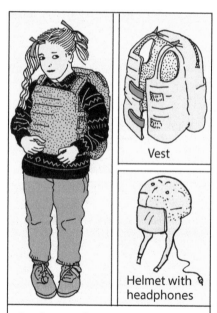

Back-to-school armor—Book-pack-vest armor and bullet-proof hats are today worn by schoolchildren.

Pedestrian outfits—Angry, fearful citizens appear on sidewalks wearing a queer melange of police shields and body armor. Fashion designers, sensing a trend, come out with wildly patterned styles in cheerful colors.

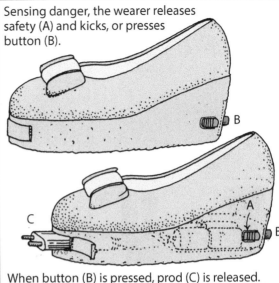

Sensing danger, the wearer releases safety (A) and kicks, or presses button (B).

When button (B) is pressed, prod (C) is released.

Shockshoes—Women say they feel safer walking on unfamiliar urban streets when wearing **Shockshoes**. These mid-heel walkers conceal a taser-like set of spring-loaded prongs that shock a potential assailant with 4,000 volts when he or she is kicked. The batteries may need to be replaced after each use.

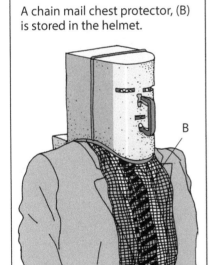

A chain mail chest protector, (B) is stored in the helmet.

Lunch is packed in a plastic bag (A) that is attached by a cord to the inside of the helmet.

Lunch Pail Helmet—The fear of a potential mugging can turn a pleasant walk from inner city office to parking lot into a nightmare. Carrying the lunch pail helmet may give one a greater sense of security.

BEING WITH NATURE

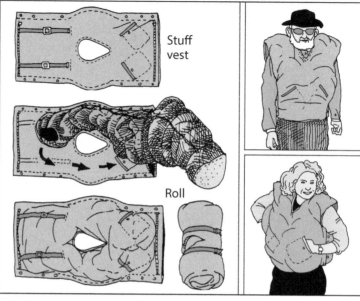

Stuff vest

Roll

Snowshoe hat—Snowshoes are similar in size and shape to some hats. The combo is very handy when walking in snow.

Sleeping bag Stuff Vest—Why carry a sleeping bag in a stuff sack when a **Sleeping Bag Stuff Vest** serves the purpose, while also providing an extra item of clothing? As temperatures drop, the vest's bulkiness is forgotten.

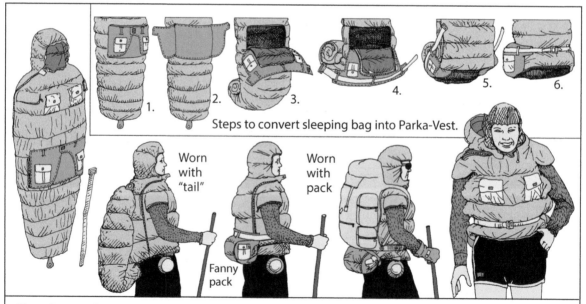

1.   2.   3.   4.   5.   6.

Steps to convert sleeping bag into Parka-Vest.

Worn with "tail"

Worn with pack

Fanny pack

Sleeping Parka-Vest—Wearing the **Sleeping Parka-Vest**, backpackers have one less item to carry during extreme weather. An early model had an exposed "tail" that was later rolled up inside a rear fanny pack.

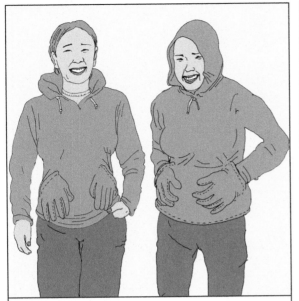

Glove Handles—Some say **Glove Handles**, attached to a hoodie in the place where pockets should be, are less than useful. At least you know you'll never lose a glove.

Elephant Man Parka —This parka allows the wearer to pre-warm their breath in bitter cold, sub-zero conditions. The face is protected by large, snap-in-place ear flaps.

Tanning hat

3-way

Duckbill

Headgear for extreme cold—The tanning hat's reflective surface folds out of the way. The three-way winter hat is easily adjusted depending on if it's sunny, chilly or stormy.

Inflatable swim wear—It takes patience to inflate this lovely swim wear, blowing on the tube before entering the water. There is an adapter for a CO_2 cartridge.

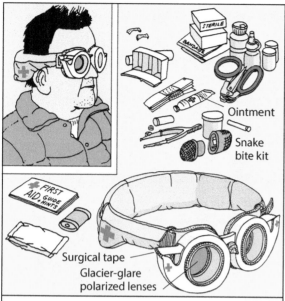

First Aid goggles—These goggles are great if you're exposed to snow and glacier glare; slippery, icy streams; loose boulders; mosquitoes; rattlesnakes; or polluted streams.

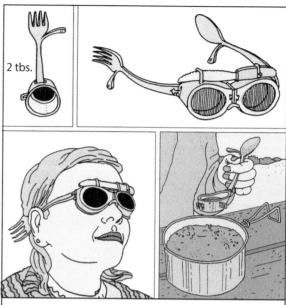

Goggles Dining Kit—This is a complete eating utensils kit for a single camper. Ladles can be used as a 2 tbs. measurer when preparing cookout recipes.

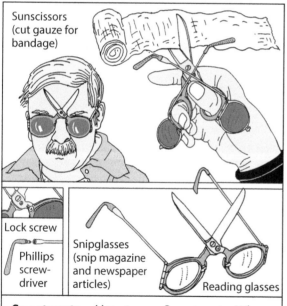

Spectacuts—How to use **Spectacuts**: When scissors are needed, unhinge the lenses and unscrew the lock screw with the Phillips screwdriver. Or wear these as glasses.

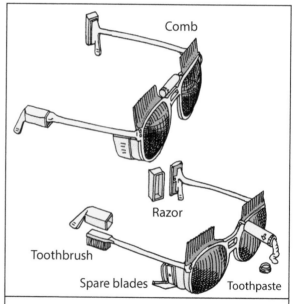

Men's Sani-glasses—For men's facial care while camping: They include comb, toothbrush, toothpaste, razor, razor blades and a mirror surface for checking shaving results.

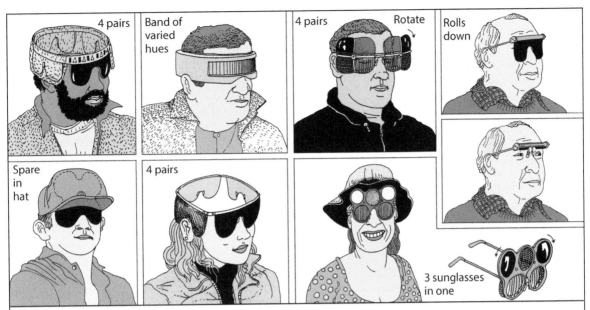

Multi-choice Glasses—When outdoors in wilderness at a high altitude, one should bring along a pair of **Multi-choice Glasses** to handle fast-changing weather conditions. These can adjust to blinding glare from polished granite or glacier ice, to a snowstorm, or to dark, brooding skies.

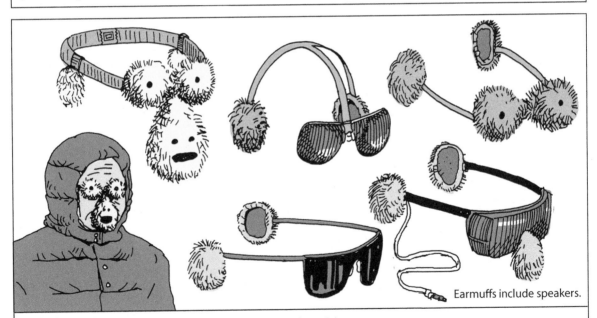

Ear, Eye and Nosemuffs—Earmuffs are a comfortable and necessary piece of head gear as severe winter weather sets in. Other types of warm muffs may be needed to help protect the eyes, nose and mouth.

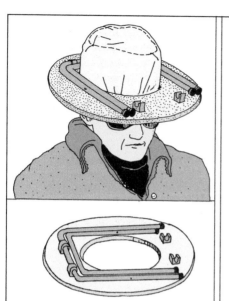

Set up for use as a toilet

Felt brim cover

Commodora—This hat "accommodates" your personal needs while you're in the wild. A fully functional camp toilet with aluminum legs is concealed inside a fedora. The brim cover hides a supply of plastic bags and toilet paper.

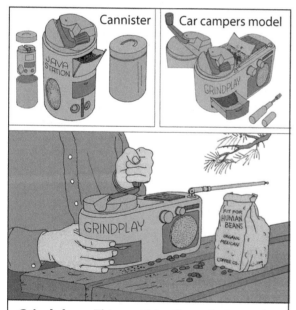

Cannister

Car campers model

Grindplay—This combination windup radio and coffee bean grinder is perfect for campers who desire freshly-ground coffee as they listen to conspiracy or End Times talk shows.

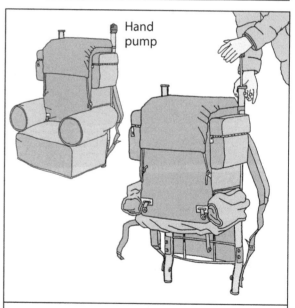

Hand pump

Inflatable chair— This backpack includes an inflatable chair! It can be inflated by breath, helium, a CO_2 cannister, air compressor, or a hand pump in the frame.

MOUNTAIN GEAR

Two-story tent—it can be strung up between two trees if they're the right distance apart. If not, this is of no use!

Umbrella tent—There's no tent anywhere in the world that is easier to set up. If you can open an umbrella, you can open this tent. In a strong gust of wind, a standard umbrella may open inside-out. That is how this tent opens!

Tent peg

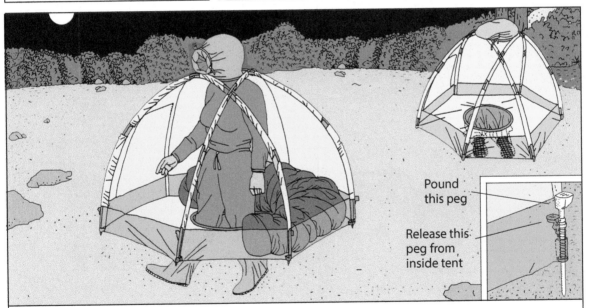

Pound this peg

Release this peg from inside tent

Four Season Insomniac—At night, campers may need to crawl out of a sleeping bag, climb out of their tent, and walk to nearby bushes or a campsite restroom to meet urgent needs. This is not necessary with the **Four Season Insomniac!** It includes a hood and waterproof walking boots!

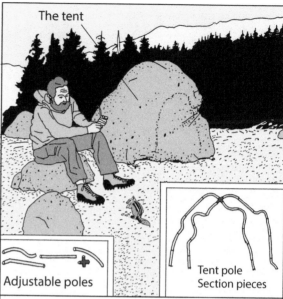

Parka Place—Backpacking equipment can be combined into a single item to save on weight. The **Parka Place** doubles as a slightly bulky jacket and a small, one-person tent.

Boulder Tent—Theft of one's gear while backpacking is a somewhat rare occurrence, yet one cannot be too careful. The boulder tent in chert, granite or sandstone, blends in.

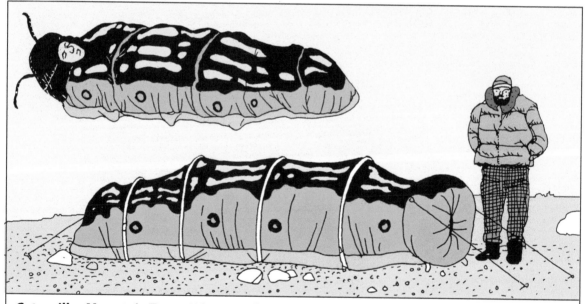

Caterpillar Mountain Tent—When you feel like you're about to freeze to death at a high-altitude campsite, do you lose your sense of humor? The **Caterpillar Mountain Tent** is designed for sophisticated mountaineers who appreciate whimsy.

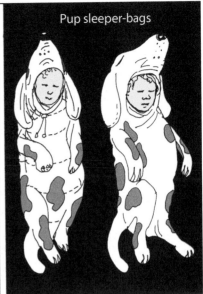

Pup sleeper-bags

Pup Tent—The original pup tent was a simple design used by armies in the field beginning in the mid-19th century, and by the Boy Scouts. This tent, with its entry snout, eyes, and spots is a visual pun. When children grow up, they will refuse to climb into their childish-looking pup tent and sleeping bags!

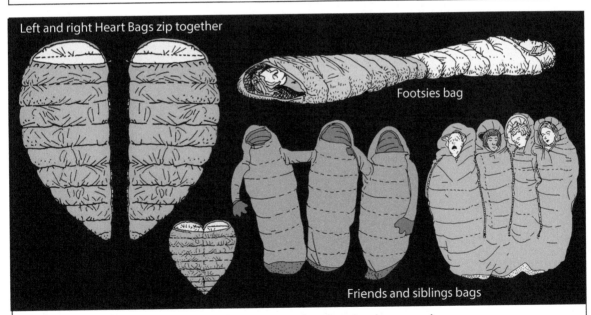

Left and right Heart Bags zip together

Footsies bag

Friends and siblings bags

Togetherness Sleeping Bags—Some rectangular sleeping bags can be zipped together to make a large, two-person bag. **Togetherness Sleeping Bags** are designed for use by more than one person. The **Heart Bag** is good for couples who have not yet broken up or gotten a divorce.

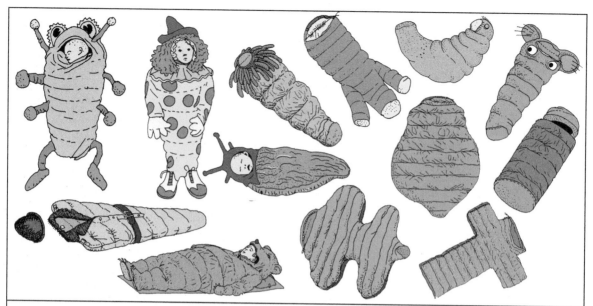

Odd-shaped sleeping bags—Amusing, but inefficient sleeping bag shapes aren't right for serious backpackers. A mummy bag should offer warmth and comfort with minimum weight. Leg room, compressibility and breathability are also factors to consider.

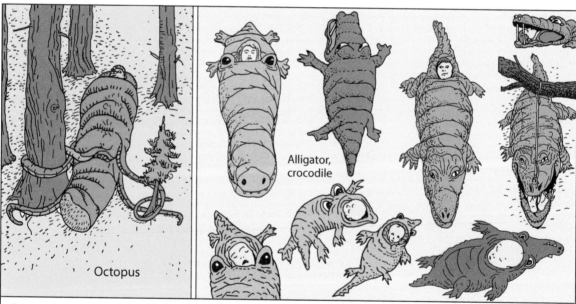

Octopus

Alligator, crocodile

Bags suggesting reptiles and sea creatures—There have been no studies to assess the effect on a child's psyche of having slept in an alligator or octopus mummy bag. To what extent does the child think it is just fun or funny, or on the other hand, terrifying?

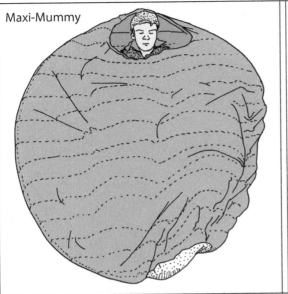

Maxi-Mummy

Whole Earth
sleeping
bag

Full Circle bags—The designer may, at times, choose a design that he/she knows is foolish or worthless. What occasionally happens is that the design finds a user and a client base. What clients might these be perfect for?

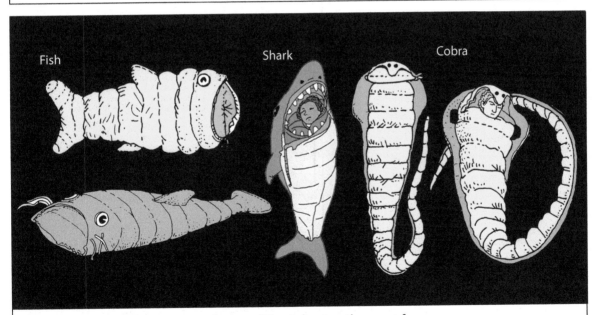

Fish

Shark

Cobra

Fish, shark and cobra mummy bags—Who is having the most fun: the designer of amusing toys and clothing items for children, or the children for whom they are designed? Is the child's shriek of joy at all similar to the designer's joy when he/she creates the object?

Midnight Mummy—Backpackers and serious campers enjoy the thermal efficiency of a mummy bag, but can't easily move or sit up once they're zipped inside. With its rudimentary flipper-like feet and arms, the **Midnight Mummy** lets one travel awkwardly and slowly, in tiny steps.

Meditator's Mountain Bag—If you are a person who's addicted to meditating regularly at night, you may find it impossible to sit up in full lotus position in cold weather while zipped inside your mummy bag. Yet it's not a problem if you own the **Meditator's Mountain Bag**!

Backpack Hotel—You may need to sleep all night in a sitting posture in the **Backpack Hotel.** The bag's inner surface is insulated with reflective material to retain body heat.

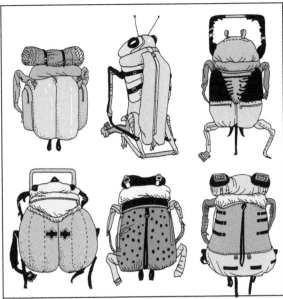

Bugpacks—These are a cult item among sophisticated backpackers who are also university entomologists. They love **Bugpacks** and also **Centipede** mummy bags.

Drollpacks—**Drollpacks** come in a range of sizes and styles that are made to meet the tastes of mountaineers as well as young children. The **Racsack** is the most popular.

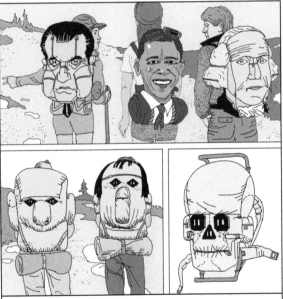

Yuckpacks—In the **Yucksacks** line, you choose a desired face style. Zippered pockets and flaps are placed to mimic facial contours. The best seller is the **Skull.**

NEW TYPES OF SPORTS

Personal fishing machine—It recommends lures and then locates, catches and nets fish, thus cutting fishing time in half!

Semi-automatic golfing machine—Even when feeling ill or lazy, a golfer may take the **Semi-Automatic Golfing Machine** to the links. The aim, angle, and force of the swing can be programmed. An on-screen caddy gives advice!

Terrain analysis

Semi-automatic tennis machine—Like the **Semi-Automatic Golfing Machine**, this device removes the exertion associated with the game, but not the mental calculations that actual players must make as they run about the court.

Moving baskets—Nets that slowly rotate, or that gently jiggle, add new challenges to basketball.

Four baskets, rotating

Calder mobile hanging nets

Power worship—A health-oriented church offers exercise in the context of a joyful, sweaty, evangelical atmosphere. Vocal prayer, lead by a physically-fit prayer leader, is timed to match the pace of stair-climber exercise machine routines.

Bonus baseball—The game combines features of pinball and baseball. Scores run high, as players get extra points for hitting the ball through rings or into pits, holes and tunnels. Runners won't be tagged out if they hide in a tunnel at first or third base.

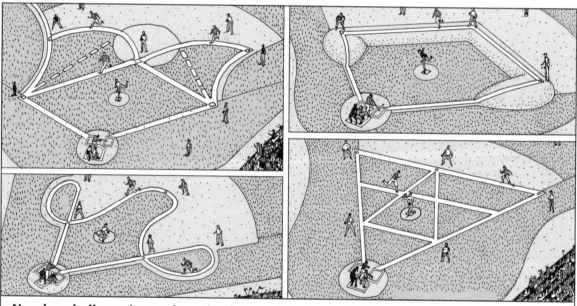

New baseball running paths—The design of the baseball field was standardized long ago with home plate, first, second, and third bases. Yet in our lives today we expect a multiplicity of decision paths and fast-changing rules. Here, a baseball field mirrors the world we live in.

PUBLIC SERVICES

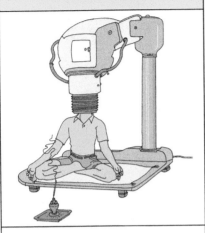

Enhanced Meditation—Machines are designed that augment natural brain waves correlated with meditation.

NYC ⓣAXI
PAY WHAT YOU CAN AFFORD

Pay what you can afford—As poverty spreads through the population new quasi-public, compassionate services offer free clothing, meals and books as well as "Pay What You Can Afford" transportation systems.

ADULT UNDERWEAR Incontinence World

BULGE CUT
TROUSERS

JODHPUR STYLE

CONFIDENCE WEAR
FEARLESS BRAND
FLUID-LOCK FEATURE
WETNESS INDICATOR

EXTRA ABSORBENT
• DISPOSABLE
• QUIET

Seniors-only malls—No one wants to talk about the next big trend. **Seniors-only malls** are filling shop spaces in formerly empty shopping centers. They feature restrooms every fifty feet, benches and grab rails, oxygen bars, breath-catching lounges, and fainting stations.

Solar cook-a-mat—At community solar cooking stations, rules are posted for properly cleaning up. Users are urged to lock glass lids with keys provided at an office.

Bumper carts—Entertainment-oriented supermarkets allow good-natured, wholesome horseplay while shopping. Electric bumper carts encourage playfulness and camaraderie.

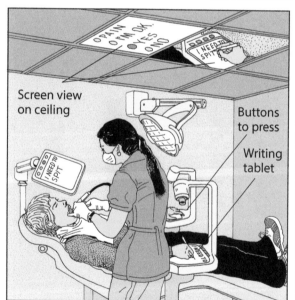

Gagtalk—With gagtalk, a patient is able to "speak" during dental work. He or she writes on a pad or presses buttons to answer a question or to express an urgent need.

Public therapy buses—People with mental or emotional upset, uncontrolled fears, or unacceptable urges may board a **public therapy bus** and speak immediately to a counselor.

Cro Magnon hotel chain—Working Americans who can't afford to fly to Bali can use these super-primitive hotel accommodations with no electrical outlets, no lighting, and no showers. A sign says "Shirt, Shoes, No Service." You grind acorns, make fire, eat weeds and kill small game for food.

THE AUTHOR

Steven M. Johnson was born in 1938 in San Rafael, California. He grew up in Berkeley and Palo Alto. He was educated at Yale University and the University of California at Berkeley. His whimsical product concepts first appeared in *The Sierra Club Bulletin* in 1973 and have since been published in numerous magazines and online publications. In 2013, he gave a TEDx talk, "Inventing Without a Purpose." In 2014, he gave a keynote speech in Orlando, Florida at the World Future Society's annual conference.

His first book, published in 1984, was "What The World Needs Now". A second and third edition of the book have been published. In 1991, "Public Therapy Buses" was published. Subsequently, there was a second edition in 2013, and a third in 2015. In 2012, "Have Fun Inventing" was published. In 2015, "Patent Depending: Vehicles" was published in color. His Web site is www.patentdepending.com.

He lives in a suburb of Sacramento, California with Beatrice, his wife of 50 years. His son, Alex S. Johnson is an author and editor who also lives in California.

CPSIA information can be obtained
at www.ICGtesting.com
Printed in the USA
FSOW04n0626110817
37333FS